Science and Beyond: Investigating the Possibility of an Afterlife

ROBERTO MIGUEL RODRIGUEZ

Copyright Page

TITLE: Science and Beyond: Investigating the Possibility of an Afterlife

1ST Edition

ISBN: 9798223229216

Table of Contents

Science and Beyond: Investigating the Possibility of an Afterlife

By Roberto Miguel Rodriguez

Chapter 1: What Happens When We Die?

The Mystery of Death

Death is an enigma that has fascinated humanity since the beginning of time. What happens when we die? Do we simply cease to exist, or is there something more beyond this life? These questions have sparked countless debates, theories, and beliefs across various cultures and religions. In this subchapter, we will delve into the intriguing topic of death and explore the possibility of an afterlife.

Near-death experiences (NDEs) have been reported by individuals who have come close to death and then been revived. These accounts often involve vivid and profound experiences, such as traveling through a tunnel of light, encountering deceased loved ones, or feeling a sense of peace and transcendence. We will explore the fascinating accounts and theories surrounding NDEs and what they might reveal about the afterlife.

Reincarnation is a belief held by many cultures that suggests the cycle of birth, death, and rebirth. We will investigate case studies and research on past-life memories, examining the possibility that our souls may continue on a journey beyond death.

Religious and spiritual perspectives on the afterlife vary greatly. Concepts like heaven, hell, and purgatory offer different interpretations of what awaits us after death. We will examine these various beliefs and the cultural and historical views that have shaped them.

Afterlife communication is a practice that has long intrigued humanity. Mediumship, séances, and channeling are methods believed to enable communication with the deceased. We will delve into these practices, exploring their history and the possibility of connecting with those who have passed away.

From a scientific standpoint, consciousness and energy hold key elements in understanding the possibility of an afterlife. We will analyze scientific research and theories on these subjects, exploring the empirical evidence that may suggest life beyond death.

Death rituals and customs vary greatly across different cultures, shedding light on their perspectives on the afterlife. We will investigate these diverse practices, burial traditions, and funeral rituals, gaining insights into how different societies understand and commemorate the transition from life to death.

The fear of death is a universal human experience that has profound psychological and philosophical implications. We will examine how this fear influences our beliefs and behaviors, and how it shapes our understanding of mortality.

Skepticism and atheism provide contrasting viewpoints to the existence of an afterlife. We will consider the arguments put forth by skeptics and atheists, exploring their perspectives and the reasons behind their rejection of life after death.

Lastly, we will explore the ethical and moral considerations that arise from our beliefs about the afterlife. How do our views on what happens when we die impact our understanding of ethics, morality, and the meaning of life?

Join us on this journey as we investigate the possibility of an afterlife, combining scientific research, cultural perspectives, and spiritual beliefs to unravel the mystery of death.

Historical Perspectives on Death

Throughout history, the question of what happens after death has captivated the human mind. From ancient civilizations to modern societies, people have sought to unravel the mysteries surrounding the afterlife. By examining historical perspectives on death, we can gain insights into how different cultures and belief systems have approached this timeless question.

In ancient Egypt, death was seen as a transition to the afterlife. The Egyptians believed in the existence of a soul that would embark on a perilous journey, requiring protection and guidance. To ensure a successful transition, they developed complex burial rituals and constructed elaborate tombs filled with treasures and offerings. These practices reflected their belief in the preservation of the body and its connection to the afterlife.

In contrast, ancient Greeks held a more philosophical view of death. Influenced by the teachings of philosophers such as Socrates and Plato, they saw death as the liberation of the soul from the physical body. According to their beliefs, the soul would continue its existence in the realm of the divine, where it would be judged for its actions in life. This perspective emphasized the importance of leading a virtuous life and cultivating wisdom.

In medieval Europe, religious beliefs played a significant role in shaping perceptions of the afterlife. Christianity introduced concepts such as heaven, hell, and purgatory. The idea of eternal reward or punishment based on one's moral conduct in life had a profound impact on the collective consciousness. Death became a moment of reckoning, prompting individuals to seek salvation through religious practices and piety.

During the Enlightenment period, scientific and philosophical thought challenged traditional beliefs about the afterlife. Skepticism and atheism gained prominence, with thinkers like David Hume questioning the existence of an immortal soul. This period saw a shift towards more secular views on death, emphasizing the finality of life and the importance of making the most of the present.

As we progress into the modern era, advancements in science and technology have further influenced our understanding of death. Near-death experiences, for instance, have sparked debates about the possibility of an afterlife. Research has explored the accounts and theories surrounding these phenomena, shedding light on the nature of consciousness and the potential existence of a realm beyond death.

By exploring historical perspectives on death, we gain a deeper appreciation for the diverse ways in which human societies have grappled with the question of what happens when we die. These perspectives highlight the profound impact that beliefs about the afterlife have on our understanding of ethics, morality, and the meaning of life. Whether rooted in religion, philosophy, or science, these perspectives continue to shape our perceptions and influence our behavior as we contemplate the mysteries that lie beyond the threshold of death.

Scientific Theories on Death and Consciousness

In the quest to understand the mysteries of life and what lies beyond, scientists have put forward various theories on death and consciousness. These theories aim to provide empirical evidence and shed light on the possibility of an afterlife. While the scientific community may not have all the answers, their research and theories offer intriguing insights into this age-old question.

One prominent theory is the concept of near-death experiences (NDEs). Exploring the accounts and theories surrounding NDEs, scientists have found that individuals who have come close to death often report similar experiences. These experiences include a sense of floating outside their bodies, encountering a bright light, and feeling a profound sense of peace and love. Some scientists argue that these experiences suggest the existence of an afterlife, while others propose that they are simply the result of neurochemical processes in the brain.

Reincarnation is another belief that has captivated the human imagination for centuries. Investigating the cycle of birth, death, and rebirth, scientists have examined case studies and research on past-life memories. Some studies have shown that children can recall detailed information about a past life that they could not have learned through normal means. While skeptics offer alternative explanations, such as cryptomnesia or fabrication, these cases continue to intrigue researchers and challenge our understanding of consciousness.

Scientific perspectives on consciousness and the possibility of an afterlife also delve into the nature of energy. Some theories propose that consciousness is not solely dependent on the physical body but can exist independently as a form of energy. This idea has led to the exploration of practices such as mediumship, séances, and channeling as possible means of communicating with the deceased.

While scientific research and theories provide valuable insights, it is essential to consider cultural and historical views on death and the afterlife. Different cultures and historical periods have interpreted and answered the question of what happens when we die in diverse ways. From the Egyptian belief in the afterlife to the Buddhist concept of reincarnation, these perspectives highlight the rich tapestry of human understanding.

The fear of death, a universal human experience, also influences our beliefs and behaviors. Examining the psychological and philosophical implications of this fear, scientists explore how it shapes our understanding of ethics, morality, and the meaning of life. Additionally, skepticism and atheism offer contrasting viewpoints, rejecting the idea of an afterlife and providing arguments against its existence.

In conclusion, scientific theories on death and consciousness provide a fascinating exploration of what happens when we die. From near-death experiences to the belief in reincarnation, these theories challenge our understanding of consciousness, energy, and the possibility of an afterlife. By analyzing scientific research, cultural perspectives, and the impact of beliefs on ethics and morality, we can gain a deeper understanding of this profound and timeless question.

Chapter 2: Near-Death Experiences: Exploring the Accounts and Theories Surrounding the Phenomenon

Defining Near-Death Experiences

Near-death experiences (NDEs) have intrigued and fascinated people for centuries. These extraordinary encounters occur when an individual comes close to death or is clinically dead but is later revived, bringing back vivid memories and profound insights. In this subchapter, we will explore the accounts and theories surrounding NDEs, shedding light on what they might reveal about the afterlife.

NDEs are often characterized by a series of remarkable elements that transcend the boundaries of ordinary human experience. Many individuals report a sense of floating out of their bodies, observing the scene from above. They describe a feeling of peace and serenity, accompanied by a tunnel of light drawing them towards an otherworldly realm. Some encounter deceased loved ones or spiritual beings, while others recall a life review, where every action and intention is laid bare.

These accounts have sparked numerous theories about the nature and significance of NDEs. Some propose that these experiences are mere hallucinations caused by oxygen deprivation or the brain's attempt to make sense of a traumatic event. Others argue that NDEs offer glimpses into an alternate reality or a spiritual dimension. Some researchers suggest that the brain releases powerful chemicals during the dying process, creating a vivid and transformative experience.

Scientific investigations into NDEs have yielded intriguing findings. Studies have shown that patients who have had NDEs often exhibit increased empathy and a reevaluation of their priorities and values. Additionally, some individuals report enhanced psychic abilities or a heightened sense of intuition following their NDE. These observations raise questions about the nature of consciousness and its potential connection to the afterlife.

While NDEs have been predominantly associated with Western cultures, similar experiences have been documented across different times and societies. These cross-cultural accounts suggest that NDEs are not merely a product of individual imagination but may hold universal significance.

In this subchapter, we will delve into the phenomenon of NDEs, exploring the various theories, scientific research, and cultural perspectives surrounding them. By examining these extraordinary experiences, we hope to deepen our understanding of the possibility of an afterlife and its implications for our lives here and now. Whether you are a skeptic, a believer, or simply curious, join us on this intriguing journey as we investigate the profound mysteries of near-death experiences.

Common Elements of Near-Death Experiences

Near-death experiences (NDEs) have fascinated and puzzled humanity for centuries. These extraordinary accounts, reported by individuals who have been on the brink of death and returned, share several common elements that transcend cultural, religious, and historical boundaries. In this subchapter, we will explore these common elements, shedding light on what they might reveal about the possibility of an afterlife.

One of the most frequently reported elements of NDEs is the sensation of leaving the physical body. Many individuals describe observing their own body from an out-of-body perspective, often floating above it. This sense of detachment from the physical form is often accompanied by a feeling of peace, tranquility, and freedom from pain.

Another common element is the presence of a bright light. NDE experiencers often describe encountering a radiant light that is warm, loving, and all-encompassing. This light is often associated with a sense of unconditional love and acceptance, and some individuals report communicating with deceased loved ones or spiritual beings within this light.

Many NDEs also involve a life review, where individuals relive significant moments from their lives. This review is often described as a non-judgmental process that allows individuals to gain insight, understanding, and a deeper sense of purpose. Some individuals report encountering a "being of light" who guides them through this review and provides wisdom and guidance.

Additionally, NDEs frequently involve encounters with deceased loved ones or spiritual beings. These encounters are often described as vivid, real, and emotionally intense. Individuals report feeling a deep sense of connection and love during these encounters, which often leave a lasting impact on their beliefs and perspectives.

Interestingly, despite the diverse cultural and religious backgrounds of NDE experiencers, the core elements of their experiences remain remarkably consistent. This suggests that NDEs may provide a glimpse into a universal, transcendent reality beyond the physical realm.

While skeptics and atheists may argue that NDEs can be explained by physiological and psychological factors, the profound and transformative nature of these experiences cannot be easily dismissed. Scientific research continues to explore the possibility that NDEs may offer valuable insights into the nature of consciousness, the afterlife, and the meaning of life itself.

In conclusion, the common elements of near-death experiences provide a compelling body of evidence that challenges our traditional understanding of death and the afterlife. These experiences offer hope, comfort, and a deeper appreciation for the mysteries of existence. By studying and understanding these accounts, we can expand our perspectives and explore the profound possibilities that lie beyond the threshold of death.

Theories on the Nature of Near-Death Experiences

Near-death experiences (NDEs) have long fascinated both scientists and the general public. These extraordinary accounts offer glimpses into what may lie beyond the threshold of death, providing valuable insights into the nature of consciousness and the possibility of an afterlife. While skeptics remain unconvinced, numerous theories have emerged to try and explain these profound and life-altering experiences.

One prevailing theory suggests that NDEs are simply hallucinations or illusions created by the brain during moments of extreme stress or trauma. According to this view, the brain releases a surge of chemicals that produce vivid and otherworldly sensations. While this theory offers a plausible explanation for some aspects of NDEs, it fails to account for the consistency and profound impact these experiences often have on individuals.

Another theory proposes that NDEs are the result of the brain's attempt to cope with the fear of death. This theory suggests that the mind creates a comforting narrative of an afterlife as a means of easing the anxiety associated with dying. While this explanation provides insight into the psychological aspects of NDEs, it does not fully explain the veridical information often reported by those who have had near-death experiences.

A more spiritual perspective posits that NDEs are glimpses into the realm of the afterlife. Proponents of this theory argue that these experiences offer evidence of an existence beyond the physical body, where consciousness continues to exist in a different form. This theory resonates with many individuals who have had NDEs, as they often describe encounters with deceased loved ones or a profound sense of peace and unconditional love.

Some researchers explore the possibility that NDEs may be a result of the brain's interaction with higher dimensions or alternate realities. This theory draws from concepts in quantum physics and posits that consciousness may transcend the physical realm and access other dimensions or planes of existence. While this theory remains highly speculative, it opens up intriguing possibilities for further exploration.

Ultimately, the mystery surrounding NDEs persists, leaving room for continued scientific inquiry and personal contemplation. As science and spirituality intersect, the search for answers about the nature of life after death continues to captivate and challenge our understanding of existence itself. By delving deeper into the phenomenon of NDEs, we may unlock profound truths about the human experience and the possibility of an afterlife.

Chapter 3: Reincarnation: Investigating the Belief in the Cycle of Birth, Death, and Rebirth

Understanding Reincarnation

Reincarnation is a fascinating and deeply rooted belief that has been embraced by cultures across the globe for centuries. It is the belief in the cycle of birth, death, and rebirth, where the soul is believed to continue its journey in a new physical body after death. This subchapter delves into the concept of reincarnation, providing insight into case studies, research on past-life memories, and the various religious and spiritual perspectives surrounding this phenomenon.

Throughout history, there have been numerous accounts of individuals who claim to have memories of past lives. These cases often involve young children who vividly recall details of people, places, and events from a time before their birth. Researchers have diligently studied these cases, documenting the evidence and attempting to explain these past-life memories. By examining the fascinating stories and experiences of individuals who claim to have lived before, we can explore the possibility of reincarnation and its implications for the nature of consciousness and the afterlife.

From a scientific perspective, reincarnation poses intriguing questions about the nature of consciousness and the continuity of the soul. This subchapter offers an analysis of scientific research and theories on consciousness, energy, and the possibility of an afterlife. By examining empirical evidence and theories, readers can gain a deeper understanding of how science intersects with the concept of reincarnation.

Moreover, this subchapter also delves into the cultural and historical views on reincarnation. Different cultures and historical periods have interpreted and answered the question of what happens when we die in diverse ways. By exploring these cultural perspectives, readers can gain a broader understanding of the beliefs and practices surrounding reincarnation.

Additionally, this subchapter provides an overview of spiritual beliefs and practices related to reincarnation, such as mediumship, séances, and channeling. These practices offer a potential means of communicating with the deceased and gaining insights into the afterlife. By delving into these practices, readers can explore the possibility of afterlife communication and its relationship to the belief in reincarnation.

Understanding reincarnation is not only an intellectual exercise but also raises ethical and moral considerations. This subchapter explores how beliefs about the afterlife impact our understanding of ethics, morality, and the meaning of life. By examining how the belief in reincarnation shapes our perspectives on life and death, readers can gain a deeper understanding of the profound influence these beliefs have on our lives.

In conclusion, this subchapter on understanding reincarnation provides a comprehensive exploration of the belief in the cycle of birth, death, and rebirth. Through case studies, research, scientific perspectives, cultural and historical views, and ethical considerations, readers can gain a deeper understanding of the complexity and significance of this belief. Whether you are a skeptic, a believer, or simply curious about the afterlife, this subchapter offers valuable insights into one of humanity's most enduring and profound beliefs.

Case Studies on Past-Life Memories

In the realm of investigating the possibility of an afterlife, the concept of reincarnation stands as a fascinating and often controversial topic. The belief in the cycle of birth, death, and rebirth has captivated the minds of individuals across cultures and historical periods. This subchapter delves into case studies and research on past-life memories, shedding light on the intriguing phenomenon of remembering previous lives.

Past-life memories refer to the recollection of events, places, or people from a previous existence. While skeptics may dismiss these accounts as mere fantasy or imagination, numerous compelling cases have emerged, challenging conventional beliefs about life and death.

One remarkable case study involves James Leininger, a young boy from Louisiana. Starting at the age of two, James began recounting vivid memories of being a World War II pilot named James Huston Jr., who was tragically killed in action. Astonishingly, the details he shared aligned with historical records, including the type of aircraft he flew and the name of the aircraft carrier he served on. Through extensive investigation, researchers were able to verify and corroborate many of James' claims, leaving them stunned by the accuracy of his past-life memories.

Another intriguing case involves Shanti Devi, an Indian girl born in 1926. From an early age, Shanti claimed to be the reincarnation of a woman named Lugdi Devi, who had died shortly after giving birth. Shanti vividly recalled details of her past life, including the names of family members and the village she had lived in. Astonishingly, when she visited her previous family, she recognized them and recounted intimate details about their lives that no one else could have known. This case garnered widespread attention and even caught the interest of Mahatma Gandhi, who deemed it worthy of investigation.

These case studies, along with many others, raise profound questions about the nature of consciousness, the continuity of the soul, and the possibility of an afterlife. While the scientific community remains divided on the interpretation of such experiences, they undoubtedly challenge our conventional understanding of life and death.

By examining these compelling accounts, we can begin to appreciate the complexity of the human experience and the mysteries that lie beyond the realm of what we currently know. Whether one chooses to believe in the existence of past-life memories or not, it is undeniable that they have the power to reshape our perspectives on life, death, and the eternal nature of the soul.

In the following chapters, we will continue exploring various aspects of the afterlife, inviting you to embark on a captivating journey that combines scientific research, cultural perspectives, and personal beliefs. Together, we will delve into the depths of human curiosity and unravel the enigma of what happens when we die.

Scientific Research on Reincarnation

Reincarnation is a belief that has captivated the human imagination for centuries. The idea that our souls continue to live on after death, inhabiting new bodies and experiencing multiple lifetimes, has been a topic of fascination and debate across cultures and religions. But is there any scientific basis to support the concept of reincarnation? In this subchapter, we will explore the scientific research on reincarnation and examine the evidence and theories surrounding this intriguing phenomenon.

One of the most fascinating aspects of reincarnation research is the investigation of past-life memories. Numerous case studies have been conducted, documenting children who vividly recall details of previous lives that they could not have possibly known. These memories often include specific names, places, and events, which are later verified to be accurate historical facts. Researchers have painstakingly collected these accounts and compared them with historical records, finding astonishing matches that defy conventional explanations.

One renowned researcher in this field is Dr. Ian Stevenson, a psychiatrist who dedicated his life to studying cases of past-life memories in children. Through meticulous investigations, Dr. Stevenson amassed an extensive collection of over 3,000 cases, providing compelling evidence for the phenomenon of reincarnation. His work has been widely recognized and respected, even by skeptics, for its scientific rigor and the credibility of the witnesses involved.

Additionally, advancements in the field of consciousness studies have shed new light on the possibility of reincarnation. Scientists are now exploring the nature of consciousness and its relationship to the physical body, opening up intriguing possibilities for understanding how consciousness may persist beyond death. The idea that consciousness is not solely dependent on the brain but may exist as a separate, non-physical entity is gaining traction among researchers.

While the scientific community remains divided on the topic, more scientists are beginning to take reincarnation seriously as a subject of study. The accumulating evidence from past-life memories, coupled with the advancements in consciousness research, is challenging the conventional materialistic view of the mind and providing a new framework for understanding the afterlife.

In conclusion, scientific research on reincarnation offers thought-provoking insights into the possibility of life after death. The verifiable cases of past-life memories and the exploration of consciousness are unraveling the mysteries surrounding reincarnation. By delving into this research, we can broaden our understanding of human existence and gain a deeper appreciation for the complex and enigmatic nature of life and death.

Chapter 4: Spiritual Beliefs: Examining Various Religious and Spiritual Perspectives

Concepts of the Afterlife in Major Religions

One of the most enduring questions that has fascinated humanity throughout history is what happens to us after we die. This subchapter, "Concepts of the Afterlife in Major Religions," aims to provide an overview of the diverse beliefs and perspectives on the afterlife that exist within different religious traditions.

Religions have long offered explanations and guidance on what lies beyond death, providing comfort, solace, and a sense of purpose to their followers. From the ancient civilizations of Egypt and Mesopotamia to the major world religions of today, countless interpretations of the afterlife have emerged.

In Christianity, the concept of the afterlife centers around the idea of heaven and hell. Christians believe that those who have lived a virtuous life will be rewarded with eternal bliss in heaven, while those who have committed sins and rejected God may face eternal damnation in hell. Additionally, the idea of purgatory is present in some Christian denominations, where souls undergo a process of purification before reaching heaven.

In Islam, the afterlife is depicted as a place of reward or punishment based on a person's deeds on Earth. Muslims believe in a paradise called Jannah, where the faithful are granted eternal happiness and fulfillment, and a hellfire called Jahannam, where the wicked are condemned to suffer.

Hinduism and Buddhism, rooted in the beliefs of reincarnation, propose a cycle of birth, death, and rebirth. According to these traditions, the quality of one's actions, or karma, determines the nature of their subsequent lives. Liberation from this cycle is seen as the ultimate goal, with Hinduism aspiring toward moksha and Buddhism seeking enlightenment, or nirvana.

Other religions, such as Judaism and Sikhism, also offer their unique perspectives on the afterlife. Judaism focuses on the importance of leading a righteous life and believes in a future Messianic Age, while Sikhism emphasizes merging with the divine and attaining liberation from the cycle of rebirth.

It is important to note that these religious beliefs are deeply ingrained in the cultural fabric of different societies, shaping rituals, funeral customs, and ethical considerations. These beliefs also influence how individuals cope with death and find meaning in life.

By exploring the concepts of the afterlife in major religions, we gain a deeper understanding of the diversity of human thought and the various ways in which people navigate questions of mortality, ethics, and the purpose of existence. Whether one finds comfort in the promise of an eternal paradise or seeks enlightenment through the cycles of rebirth, the exploration of these beliefs offers a rich tapestry of perspectives on what lies beyond the realm of the living.

Heaven, Hell, and Purgatory

In the realm of spiritual beliefs, few concepts capture the imagination quite like the ideas of heaven, hell, and purgatory. These notions have been deeply ingrained in various religious and cultural traditions throughout history. They offer a glimpse into what might await us after death, fueling our curiosity about the nature of the afterlife.

Heaven, often depicted as a realm of eternal bliss and happiness, represents the ultimate reward for a life well-lived. Different religious traditions offer their own interpretations of this celestial paradise, but common themes include reunion with loved ones, communion with a divine presence, and the absence of suffering or pain. Exploring these diverse perspectives sheds light on what people hope to experience in the afterlife and how their beliefs shape their moral and ethical choices in the present.

On the other hand, hell conjures images of eternal punishment and damnation. It represents the consequences of a life marred by sin or evil deeds. While some traditions portray hell as a fiery inferno, others describe it as a place of spiritual torment or separation from the divine. The concept of hell raises questions about justice, morality, and the nature of punishment, providing a framework for understanding the consequences of our actions in this life and the next.

Between these two extremes lies purgatory, a concept that exists in certain Christian traditions. Purgatory is believed to be a temporary state of purification for souls who have not attained the perfection required for entry into heaven. Souls in purgatory undergo a process of purification, often through suffering, in order to be deemed worthy of entering heaven. This notion offers hope for those who believe in the possibility of redemption and the potential for growth even after death.

While heaven, hell, and purgatory are deeply rooted in religious and spiritual beliefs, they have also been subject to scrutiny and interpretation from a scientific and empirical standpoint. Some argue that these concepts are merely psychological constructs, reflecting our desires for justice, happiness, and meaning. Others explore near-death experiences and afterlife communication as potential evidence for the existence of these realms.

Understanding the various interpretations of heaven, hell, and purgatory provides insight into how different cultures and historical periods have grappled with the question of what happens after death. It also highlights the ethical and moral considerations that arise from these beliefs, shaping our understanding of how we should live our lives and the impact our actions have on our ultimate destiny.

In "Science and the Beyond: Investigating the Possibility of an Afterlife," this subchapter delves into the rich tapestry of beliefs surrounding heaven, hell, and purgatory, exploring their origins, cultural significance, and impact on our understanding of life, death, and the meaning of existence. By examining these concepts through multiple lenses, we hope to shed light on the profound human fascination with the afterlife and its profound influence on our beliefs, behaviors, and the way we navigate the world.

Cultural Variations in Spiritual Beliefs

The concept of the afterlife is deeply ingrained in human culture, and yet, as we explore different societies and belief systems, we find a remarkable diversity in how people envision what happens when we die. This subchapter will delve into the cultural variations in spiritual beliefs surrounding the afterlife, shedding light on the rich tapestry of human imagination and understanding.

Throughout history, different cultures have developed unique perspectives on the afterlife, shaped by their religious, philosophical, and cultural traditions. From ancient civilizations to modern societies, each has its own interpretation of what lies beyond death.

In ancient Egypt, for example, the belief in an afterlife was so central that elaborate burial practices, such as mummification, were developed to preserve the body for the journey to the next realm. The Egyptians believed in a complex underworld, where the deceased would undergo a series of trials before reaching the eternal paradise of the Field of Reeds.

In contrast, many Native American tribes hold a belief in ancestral spirits and interconnectedness with nature. They see death as a continuation of life, with the spirits of the departed remaining present and guiding the living. These cultures often emphasize the importance of honoring ancestors through rituals and ceremonies.

Meanwhile, in Hinduism, the cycle of birth, death, and rebirth, known as reincarnation, plays a central role. Hindus believe that one's actions in this life determine their fate in the next, with the ultimate goal being liberation from the cycle of rebirth and union with the divine.

These examples represent just a fraction of the vast array of spiritual beliefs found around the world. From concepts like heaven and hell in Christianity to the idea of spiritual evolution in Buddhism, each culture offers its own unique perspective on the afterlife.

Understanding these cultural variations in spiritual beliefs not only expands our knowledge but also fosters a sense of respect and appreciation for the diversity of human experience. It allows us to recognize that there is no singular truth about the afterlife, but rather a multitude of interpretations shaped by different cultural, historical, and religious contexts.

As we continue our exploration of the possibility of an afterlife, it is crucial to consider these cultural variations and their impact on our understanding of what happens when we die. By examining different perspectives, we can gain a deeper appreciation for the complexities of human spirituality and the ways in which it shapes our beliefs, rituals, and customs surrounding death.

Ultimately, this subchapter serves as a reminder that the question of what happens when we die is not a simple one. It is a question that elicits a wide range of answers, reflecting the depth and diversity of human thought and imagination.

Chapter 5: Afterlife Communication: Exploring the Possibility of Communicating with the Deceased

Mediumship: Connecting with Spirits

Mediumship is a practice that has intrigued and captivated people for centuries. It is the belief and ability to communicate with spirits of the deceased. In this subchapter, we will delve into the world of mediumship and explore the possibility of connecting with spirits from beyond the grave.

Mediumship has long been a subject of fascination and controversy. While some view it as a genuine means of connecting with loved ones who have passed away, others dismiss it as mere trickery or wishful thinking. However, with advancements in scientific research and the increasing number of personal testimonies, mediumship is gaining recognition as a legitimate avenue for afterlife communication.

There are various methods employed by mediums to establish contact with spirits. Some use tools such as tarot cards or crystal balls, while others rely solely on their intuition and psychic abilities. The process of mediumship involves the medium attuning themselves to the spirit world, allowing messages and information to flow through them from the other side.

Many individuals seek out mediums in the hopes of receiving closure or guidance from departed loved ones. Mediums claim to convey messages, memories, and emotions from the deceased, providing comfort and reassurance to those left behind. These encounters can offer a sense of connection and solace, helping individuals navigate the grieving process and find peace in knowing their loved ones continue to exist in some form.

However, it is essential to approach mediumship with an open mind and critical thinking. While there are undoubtedly gifted mediums who possess a genuine ability to connect with spirits, there are also those who may take advantage of vulnerable individuals or offer false hope. It is important to research and seek recommendations when considering consulting a medium.

In this subchapter, we will explore the history and cultural significance of mediumship, including its role in various religions and spiritual practices. We will examine case studies and research on the subject, shedding light on the potential validity and limitations of afterlife communication.

Mediumship offers a unique perspective on the afterlife, bridging the gap between the physical and spiritual realms. Whether one believes in the existence of an afterlife or approaches mediumship with skepticism, it is undeniable that the practice has a profound impact on those who seek solace and understanding in the face of loss. By exploring the world of mediumship, we can gain insight into the human desire to connect with the departed and the ongoing exploration of what lies beyond our physical existence.

Séances: Contacting the Other Side

Séances have long been a source of fascination and intrigue, offering the possibility of communicating with the deceased. These gatherings, often led by a medium, provide individuals with a platform to connect with loved ones who have passed away. In this subchapter, we will delve into the world of séances and explore the potential for afterlife communication.

Throughout history, séances have been conducted in various forms and across different cultures. The purpose of these gatherings is to bridge the gap between the living and the dead, allowing for messages and insights to be shared. Mediums, individuals who claim to have the ability to communicate with the spirit world, play a significant role in facilitating these connections.

While skeptics may dismiss séances as mere illusions or tricks, many individuals have reported profound experiences during these sessions. The séance room is typically dimly lit, creating an atmosphere conducive to spiritual communication. Participants join hands, forming a circle, and the medium acts as a conduit between the living and the deceased.

Messages from the other side can take various forms, including voices, visions, or even physical manifestations. Loved ones may offer guidance, comfort, or closure to those seeking connection. For some, the experience is transformative, providing a sense of peace and reassurance that life continues beyond death.

However, it is important to approach séances with a critical eye. While there are undoubtedly genuine mediums who possess the ability to communicate with the deceased, there are also those who exploit vulnerable individuals for personal gain. It is crucial to exercise discernment and seek out reputable mediums who operate with integrity and ethics.

In this subchapter, we will explore the history of séances, from their origins in spiritualism to their continued practice today. We will examine case studies and accounts of séance experiences, highlighting both the successes and the controversies surrounding this form of afterlife communication. By examining séances through a scientific lens, we can assess the validity and potential limitations of these practices.

Séances offer a unique perspective on the possibility of an afterlife and provide individuals with an opportunity to connect with their departed loved ones. Through an exploration of séances, we can gain a deeper understanding of the human desire to communicate with the other side and the implications this has on our beliefs, behaviors, and quest for meaning in life.

Channeling: Receiving Messages from the Beyond

In the realm of exploring the possibility of an afterlife, one intriguing avenue to consider is the practice of channeling. Channeling is the process of receiving messages or information from entities or beings from the beyond, whether they are deceased individuals, spirit guides, or higher beings. It is a fascinating phenomenon that has captured the attention of many individuals seeking answers about what happens when we die.

Channeling has been practiced for centuries across different cultures and spiritual traditions. It involves a person, known as a channeler or medium, who acts as a bridge between the physical world and the spiritual realm. Through various techniques such as meditation, trance states, or automatic writing, the channeler opens themselves up to receive messages from the beyond.

The messages received through channeling can provide insights, guidance, and comfort to individuals seeking to connect with their departed loved ones or gain a deeper understanding of the afterlife. These messages often carry personal and profound meaning for those who receive them, offering a sense of closure, reassurance, or spiritual growth.

While skeptics may question the authenticity of channeling, there have been numerous accounts of accurate and verifiable information being channeled. Some channelers have even been subjected to scientific tests and have produced remarkable results that challenge conventional beliefs about consciousness and the limits of human perception.

It is important to approach channeling with an open mind and discernment. Not all channelers are genuine, and there are certainly instances of fraud and deception. However, there are also reputable individuals who have dedicated their lives to honing their abilities and using them for the betterment of others.

In the subchapter on channeling, we will delve into the fascinating world of afterlife communication. We will explore the history and cultural significance of channeling, examining its role in different religious and spiritual traditions. We will also delve into the scientific perspectives on channeling, analyzing research and theories on consciousness and energy that provide a more empirical viewpoint.

Furthermore, we will discuss the ethical and moral considerations associated with channeling, as beliefs about the afterlife can profoundly impact our understanding of ethics, morality, and the meaning of life. We will also address skepticism and atheism's viewpoints, acknowledging the arguments against the existence of an afterlife and how they shape our understanding of death and beyond.

Channeling offers a unique perspective on what happens when we die, providing a potential avenue for communication with the deceased and insights into the mysteries of the afterlife. By exploring this practice, we hope to shed light on the possibility of a continued existence beyond death and encourage a deeper understanding of the human experience.

Chapter 6: Scientific Perspectives: Analyzing Research and Theories on the Afterlife

Consciousness and its Relation to the Afterlife

When pondering the mysteries of life and death, one cannot help but question the nature of consciousness and its connection to the afterlife. This subchapter delves into the intricate interplay between consciousness and the possibility of an afterlife, seeking to shed light on this enigmatic topic.

Throughout history, numerous accounts and theories have emerged surrounding near-death experiences (NDEs) – those fleeting moments when individuals come close to the brink of death and then return to life. These experiences offer a unique glimpse into what might await us beyond the veil. By exploring the accounts of those who have undergone NDEs, we may begin to unravel the secrets of the afterlife and gain insights into its nature.

Another intriguing concept related to consciousness and the afterlife is reincarnation. Investigating the belief in the cycle of birth, death, and rebirth, this subchapter presents compelling case studies and research on past-life memories. By examining these cases, we can question whether consciousness transcends the physical body and continues its journey through multiple lifetimes.

Moreover, an exploration of various religious and spiritual perspectives on what happens after death is essential. Concepts such as heaven, hell, and purgatory provide diverse interpretations of the afterlife. By delving into these spiritual beliefs, we can gain a deeper understanding of the human quest for eternal existence and the impact these beliefs have on our lives.

Intriguingly, some individuals claim to have communicated with the deceased through practices such as mediumship, séances, and channeling. This subchapter investigates the possibility of afterlife communication, scrutinizing these practices and their potential to bridge the gap between the living and the departed.

To approach the topic from a more empirical standpoint, scientific research and theories on consciousness and energy are also analyzed. By examining the scientific perspective on the afterlife, we can explore the possibilities beyond mere belief and delve into the realm of evidence and observation.

Furthermore, this subchapter explores how different cultures and historical periods have interpreted and answered the question of what happens when we die. By examining cultural and historical views, we can appreciate the rich tapestry of human beliefs and gain insights into the diverse ways people have sought to understand the mysteries of existence.

Death rituals and customs from around the world are also investigated, shedding light on cultural perspectives on the afterlife. By understanding how different societies honor and remember the deceased, we can gain a deeper appreciation for the various ways in which people cope with mortality and seek solace in the thought of an afterlife.

Examining the psychological and philosophical implications of humanity's fear of death is another crucial aspect of this subchapter. By understanding how our fear of death influences our beliefs and behaviors, we can begin to unravel the complex relationship between mortality and the search for meaning.

In contrast to those who believe in an afterlife, skeptics and atheists reject the idea altogether. This subchapter considers their viewpoints and arguments against the existence of life after death, providing a balanced perspective on the topic.

Finally, the subchapter explores the ethical and moral considerations stemming from beliefs about the afterlife. By analyzing how these beliefs impact our understanding of ethics, morality, and the meaning of life, we can gain a deeper understanding of the profound influence the afterlife has on our daily lives.

In conclusion, this subchapter on consciousness and its relation to the afterlife delves into the profound questions surrounding our existence beyond death. By exploring near-death experiences, reincarnation, spiritual beliefs, afterlife communication, scientific perspectives, cultural and historical views, death rituals, fear of death, skepticism, atheism, and ethical considerations, we aim to provide a comprehensive exploration of the possibilities and implications of an afterlife.

Energy and its Role in the Afterlife

When contemplating the mysteries of the afterlife, one cannot ignore the concept of energy. Energy, in its various forms, plays a crucial role in understanding what happens when we die. Whether it is the energy that sustains our physical bodies or the metaphysical energy that transcends our earthly existence, exploring this fundamental force can shed light on the possibilities of an afterlife.

In the realm of near-death experiences (NDEs), individuals often report encountering a profound sense of energy. These accounts provide valuable insights into the nature of consciousness beyond the physical realm. Many describe a feeling of being surrounded by an intense, loving energy or being drawn towards a bright light. These experiences suggest that consciousness continues to exist beyond the confines of our physical bodies, supported by a powerful energy that transcends our current understanding.

Reincarnation, the belief in the cycle of birth, death, and rebirth, also relies on the concept of energy. Advocates argue that the energy of our consciousness carries on from one life to the next, shaping our experiences and memories. Fascinating case studies and research on past-life memories further support the notion that energy is an integral component of the afterlife.

From a scientific perspective, the study of consciousness and energy offers intriguing possibilities. Researchers have explored the idea that consciousness may be a form of energy that exists independently of our physical bodies. This hypothesis opens up new avenues for investigating the afterlife empirically, bridging the gap between science and spirituality.

Cultural and historical viewpoints also provide intriguing insights into the role of energy in the afterlife. Different cultures and historical periods have interpreted energy in distinct ways, with concepts like chi, prana, or the soul representing the life force that continues after death. These beliefs highlight the universality of energy as a fundamental aspect of the afterlife across various societies and time periods.

Furthermore, the practices of afterlife communication, such as mediumship and channeling, rely on the transfer of energy between the living and the deceased. Mediums claim to tap into this energetic connection to communicate with spirits, offering comfort and closure to those seeking contact with their departed loved ones.

In conclusion, energy is a key element in understanding the afterlife. From near-death experiences to reincarnation, scientific theories to cultural beliefs, the concept of energy permeates our exploration of what happens when we die. By delving deeper into the role of energy, we can gain a richer understanding of the possibilities that await us beyond this earthly existence.

Empirical Evidence for an Afterlife

In this subchapter, we will explore the empirical evidence that supports the existence of an afterlife. While the concept of an afterlife has long been a topic of debate and speculation, there are several intriguing scientific studies and theories that provide compelling evidence for the continuation of consciousness beyond death.

One area of research that has shed light on the possibility of an afterlife is near-death experiences (NDEs). NDEs occur when individuals are pronounced clinically dead but are later revived, often reporting remarkable experiences such as out-of-body sensations, a tunnel of light, and encounters with deceased loved ones. These accounts are consistent across cultures and have been documented in numerous studies, suggesting that they may provide glimpses into the afterlife.

Furthermore, the phenomenon of reincarnation has also been extensively studied. Researchers have documented cases of individuals who claim to remember past lives, often providing detailed and accurate information about people, places, and events that they could not have known otherwise. These cases, known as past-life memories, have been investigated by renowned scientists and have led to compelling evidence for the existence of an afterlife.

Additionally, we will explore various religious and spiritual perspectives on what happens after death. Concepts such as heaven, hell, and purgatory have been deeply ingrained in human culture for centuries, and understanding these beliefs can provide valuable insights into the afterlife.

Furthermore, we will delve into practices such as mediumship, séances, and channeling, which offer the possibility of communicating with the deceased. While skeptics may dismiss these practices as mere illusions or trickery, there have been numerous well-documented cases that defy conventional explanations, suggesting the potential for genuine afterlife communication.

From a scientific standpoint, we will analyze research and theories on consciousness and energy that provide a more empirical perspective on the possibility of an afterlife. Advances in quantum physics and the study of consciousness have challenged traditional materialistic views and opened up new avenues for understanding the nature of existence beyond death.

By examining the cultural and historical views on the afterlife, we will gain a deeper appreciation for the diverse interpretations of what happens when we die. Different cultures and historical periods have offered unique perspectives on the afterlife, enriching our understanding of this profound mystery.

Furthermore, we will explore how beliefs about the afterlife impact our understanding of ethics, morality, and the meaning of life. Understanding the ethical and moral considerations surrounding the afterlife can provide valuable insights into the human experience and the choices we make in our lives.

In conclusion, the empirical evidence for an afterlife is compelling and diverse. Through the exploration of near-death experiences, reincarnation, spiritual beliefs, afterlife communication, scientific perspectives, cultural and historical views, death rituals, fear of death, skepticism and atheism, and ethical and moral considerations, we will embark on a fascinating journey that challenges our preconceived notions and invites us to contemplate the possibility of life beyond death.

Chapter 7: Cultural and Historical Views: Exploring Interpretations of the Afterlife

Ancient Civilizations and the Afterlife

Throughout human history, the concept of the afterlife has captivated the minds of people from all walks of life. Ancient civilizations, in particular, had unique and fascinating beliefs about what happens when we die. Exploring these beliefs provides us with valuable insights into the human quest for understanding the mysteries of life and death.

In ancient Egypt, for example, the afterlife played a central role in their culture. The Egyptians believed in a complex system where the soul would journey through various stages, ultimately leading to eternal life. They placed great importance on preserving the body through mummification, as they believed the physical form was necessary for the soul to continue its existence in the afterlife.

Similarly, the ancient Greeks held a strong belief in the afterlife. Their views were influenced by the mythical stories of Hades, the realm of the dead. According to Greek mythology, after death, the soul would cross the river Styx and be judged by Hades. Depending on their actions in life, souls would either be rewarded in the Elysian Fields or condemned to eternal punishment.

In contrast, the ancient Norse civilization had a different perspective on the afterlife. They believed in a realm called Valhalla, reserved for warriors who died honorably in battle. This illustrates how cultural and historical contexts shape our beliefs about what happens after death.

The study of ancient civilizations and their beliefs about the afterlife is not merely an exercise in historical curiosity. It serves as a reminder of the universal human desire to understand what lies beyond our mortal existence. By examining these beliefs, we gain a broader perspective on the diverse ways in which different cultures have grappled with the mystery of death.

Furthermore, understanding ancient civilizations' views on the afterlife can shed light on contemporary beliefs and practices. Many spiritual and religious traditions today draw inspiration from these ancient beliefs, seeking guidance and solace in the wisdom of the past.

In conclusion, the exploration of ancient civilizations' beliefs about the afterlife provides a fascinating glimpse into humanity's enduring quest for answers about what happens when we die. By studying these beliefs, we can gain a deeper understanding of our own cultural perspectives and the profound impact they have on our lives. From ancient Egypt to the Norse Vikings, the ancient world offers a rich tapestry of ideas and beliefs that continue to shape our understanding of the afterlife today.

Cultural Perspectives on Death and the Afterlife

Throughout history, various cultures have developed unique and diverse perspectives on death and the afterlife. These beliefs have shaped rituals, customs, and spiritual practices that provide comfort and guidance to individuals as they contemplate what happens when we die.

One of the most intriguing aspects of cultural perspectives on death is the concept of the afterlife. Many religions, such as Christianity, Islam, and Hinduism, embrace the idea of an afterlife where the soul continues to exist in a different realm. These beliefs often include notions of heaven, hell, and purgatory, each with its own set of rewards or punishments. Exploring these ideas can offer insight into the values and moral codes that guide these cultures.

Another fascinating topic within the realm of cultural perspectives on death is reincarnation. Investigating the belief in the cycle of birth, death, and rebirth reveals a rich tapestry of stories and case studies. Some individuals claim to have past-life memories, providing intriguing evidence for the existence of a soul that persists beyond death.

Beyond religious and spiritual beliefs, cultural perspectives on the afterlife also manifest in death rituals and customs. Every culture has its unique way of mourning and honoring the deceased. From elaborate funeral processions to burial practices that reflect a society's values and traditions, these rituals shed light on how different cultures perceive death and the journey beyond.

However, not all cultural perspectives embrace the idea of an afterlife. Skepticism and atheism provide an alternative viewpoint, rejecting the existence of life after death. These viewpoints often rely on scientific perspectives and empirical evidence to challenge the notion of an afterlife.

Exploring cultural perspectives on death and the afterlife also invites us to examine the fear of death. This universal fear has profound psychological and philosophical implications, influencing our beliefs, behaviors, and even our understanding of ethics and morality. Understanding how different cultures navigate this fear can offer valuable insights into our own perspectives.

In conclusion, cultural perspectives on death and the afterlife provide a rich tapestry of beliefs, rituals, and customs that offer solace and guidance to individuals grappling with the question of what happens when we die. By examining these perspectives, we can gain a deeper understanding of the human experience and how our beliefs shape our understanding of life, death, and the meaning we attribute to it.

Historical Shifts in Beliefs about the Afterlife

Throughout history, human beings have grappled with the mysteries of what happens when we die. The concept of an afterlife, or some form of existence beyond death, has been a central theme in various cultures, religions, and philosophical traditions. This subchapter explores the historical shifts in beliefs about the afterlife, shedding light on how our understanding of this topic has evolved over time.

In ancient civilizations such as Egypt and Mesopotamia, the afterlife was often seen as a continuation of earthly life. Egyptians, for example, believed in the existence of a complex underworld where the soul would undergo a series of tests before entering the eternal realm. Similarly, the Mesopotamians believed in an underworld ruled by a god of death.

With the rise of major religions such as Judaism, Christianity, and Islam, new conceptions of the afterlife emerged. These monotheistic faiths introduced the notion of a judgment day, where souls would be rewarded or punished based on their actions in life. In Christianity, heaven, hell, and purgatory became central components of the afterlife, with eternal salvation or damnation awaiting individuals.

During the Enlightenment period, a shift towards more rational and secular thinking challenged religious beliefs about the afterlife. Philosophers like Immanuel Kant argued that the existence of an afterlife could not be proven through empirical evidence and instead emphasized the importance of leading a moral life in the present.

In the 19th and 20th centuries, spiritualism gained popularity, leading to a renewed interest in afterlife communication. Mediumship, séances, and channeling became common practices for those seeking to connect with deceased loved ones. This spiritualist movement challenged traditional religious beliefs and opened up new avenues for exploring the afterlife.

In recent years, scientific perspectives on consciousness and energy have influenced our understanding of the afterlife. Near-death experiences, where individuals report vivid encounters during clinical death, have sparked debates about the nature of consciousness and the possibility of an afterlife. Research on past-life memories and reincarnation has also raised intriguing questions about the continuity of consciousness beyond death.

As we delve into the historical shifts in beliefs about the afterlife, it is important to recognize that our understanding of this topic is deeply influenced by cultural, religious, and philosophical contexts. By exploring the diverse perspectives and theories surrounding the afterlife, we can gain a deeper appreciation for the complexity and mystery of what lies beyond death.

Chapter 8: Death Rituals and Customs: Investigating Diverse Practices and Traditions

Burial Practices from Around the World

In our quest to understand what happens when we die, it is essential to examine the diverse burial practices and funeral traditions from around the world. These practices shed light on cultural perspectives on the afterlife and provide valuable insights into the beliefs and rituals surrounding death.

From the ancient Egyptians' elaborate process of mummification to the Tibetan sky burials, where bodies are exposed to the elements and consumed by vultures, burial practices vary greatly across different cultures and historical periods. Each tradition is rooted in unique beliefs and customs that reflect the values and spiritual perspectives of the people.

For instance, in many indigenous cultures, such as the Native Americans and the Aboriginal Australians, burial rituals involve returning the body to the earth, symbolizing the cyclical nature of life and death. These practices emphasize the interconnectedness of all living beings and the importance of maintaining harmony with nature.

In contrast, some cultures, like the ancient Greeks and Romans, believed in an afterlife where the soul would continue its journey. They practiced cremation to release the soul from the physical body, allowing it to ascend to the realm of the divine.

Beyond cultural variations, burial practices also evolve over time. In recent years, alternative methods such as green burials and bio-cremation have gained popularity. Green burials focus on environmental sustainability, using biodegradable materials and avoiding embalming chemicals. Bio-cremation, also known as alkaline hydrolysis, uses a water-based process to reduce the body to its basic elements.

Studying burial practices not only provides us with a glimpse into different cultural and historical perspectives on the afterlife, but it also raises intriguing questions about the universality of our beliefs and the human need for rituals surrounding death.

By exploring burial practices from around the world, we can deepen our understanding of how different societies cope with the mysteries of death and what it means for our own existence. These diverse practices offer a rich tapestry of human experiences and beliefs, reminding us that the search for answers about the afterlife is a deeply ingrained part of our shared human experience.

Funeral Customs and Rituals

Funeral customs and rituals play a significant role in how different cultures and societies navigate the journey from life to death and beyond. These practices provide comfort, closure, and a sense of community during the grieving process. Understanding the diverse funeral customs and rituals from around the world can shed light on cultural perspectives on the afterlife and offer insights into the human experience of death.

In many cultures, funerals are not only a time to mourn the loss of a loved one but also an opportunity to celebrate their life. For example, in Mexico, the Day of the Dead is a vibrant celebration where families gather to honor and remember their deceased ancestors. Graves are adorned with flowers, candles, and offerings, creating a colorful and festive atmosphere. This tradition reflects a belief in the continued connection between the living and the dead.

In contrast, some cultures have more somber funeral customs. In parts of Asia, particularly in countries like China and Japan, Buddhist and

Taoist traditions influence funeral rituals. These rituals often involve elaborate ceremonies, chanting, and prayers to guide the deceased into the afterlife. The focus is on ensuring a peaceful transition and offering support to the deceased's spirit.

Furthermore, burial practices vary widely across different cultures. While some opt for traditional burials, others choose cremation as a means of disposing of the body. In Hinduism, cremation is considered a sacred act that releases the soul from the physical body, allowing it to move on to the next life. Meanwhile, in certain African cultures, burial sites hold great significance and are believed to be places where communication with the deceased is possible.

These customs and rituals demonstrate the deep connection between culture, spirituality, and the afterlife. They provide comfort to the bereaved and offer a way to honor and remember those who have passed away. By examining these practices, we can gain a deeper understanding of how different cultures perceive death and the afterlife.

In this subchapter, we will explore a variety of funeral customs and rituals from around the world. We will delve into the symbolism behind these practices and examine the cultural beliefs that underpin them. By doing so, we hope to broaden our understanding of the human experience and the myriad ways in which we cope with the mysteries of life, death, and what lies beyond.

Symbolism in Death Rituals

Death is a universal experience, and throughout history, cultures all over the world have developed unique rituals and customs to honor and remember the deceased. These death rituals not only serve practical purposes, such as disposing of the body, but also hold deep symbolic meanings, reflecting the beliefs, values, and cultural identities of the people who practice them.

One common symbolism found in death rituals is the concept of transition and rebirth. Many cultures believe in an afterlife or some form of continuation of the soul, and their death rituals often incorporate symbols and actions that represent this belief. For example, in ancient Egypt, the mummification process aimed to preserve the body for the journey to the afterlife. The elaborate rituals involved in mummification symbolized the transition from death to the eternal life that awaited the deceased.

Another symbolic element in death rituals is the notion of purification and spiritual cleansing. In Hinduism, for instance, the ritual of cremation is seen as a way to purify the soul and release it from the physical body, allowing it to continue its journey. The act of cremation symbolizes the transformation of the body into ashes, representing the impermanence of life and the soul's liberation from the material world.

Symbolism in death rituals can also be seen in the objects and artifacts used during the ceremonies. In many cultures, flowers are commonly associated with death and are used in funeral rituals to symbolize beauty, fragility, and the cycle of life. Similarly, candles are often lit to symbolize the presence of the soul and to guide it on its journey.

The symbolism in death rituals not only provides comfort and closure for the bereaved but also serves as a way to communicate and connect with the deceased. Many cultures believe that through these rituals, they can maintain a spiritual connection with their loved ones, ensuring their well-being in the afterlife. Practices such as ancestor worship, mediumship, and channeling are examples of afterlife communication methods that allow individuals to seek guidance and communicate with the departed.

Understanding the symbolism in death rituals allows us to gain insights into different cultural perspectives on the afterlife and the meaning of death. It reminds us that death is not just an end, but a transition, and that our beliefs and rituals surrounding death can shape our understanding of life, ethics, and the meaning of our existence.

In the following chapters, we will explore the diverse death rituals and customs from around the world, shedding light on the cultural perspectives on the afterlife. We will delve into the psychological and philosophical implications of our fear of death and how it influences our beliefs and behaviors. Moreover, we will analyze the scientific research and theories on consciousness, energy, and the possibility of an afterlife from an empirical standpoint.

By examining the symbolism in death rituals, we can gain a deeper understanding of the human experience and the profound questions surrounding what happens when we die.

Chapter 9: Fear of Death: Examining Psychological and Philosophical Implications

The Fear of Death and Existential Anxiety

Death is a topic that has fascinated and frightened humanity since the beginning of time. The fear of death and existential anxiety are deeply ingrained in our collective consciousness, shaping our beliefs, behaviors, and even our understanding of the meaning of life. In this subchapter, we will delve into the psychological and philosophical implications of humanity's fear of death and how it influences our perspectives on the afterlife.

For many individuals, the fear of death stems from the unknown. What happens when we die? Is there an afterlife? These questions have puzzled philosophers, scientists, and theologians for centuries. Near-death experiences provide a glimpse into this enigmatic realm, offering accounts and theories that hint at the possibility of life after death. We will explore these experiences and their potential implications for our understanding of the afterlife.

Reincarnation, the belief in the cycle of birth, death, and rebirth, is another concept that has captivated human imagination for centuries. Through case studies and research on past-life memories, we will investigate the validity of this belief and its significance in shaping our understanding of the afterlife.

Religious and spiritual perspectives play a crucial role in influencing our perceptions of what happens after death. Concepts like heaven, hell, and purgatory offer hope, judgment, and redemption. We will examine various religious and spiritual beliefs surrounding the afterlife and how they impact our ethical and moral considerations.

Additionally, we will delve into the practices of afterlife communication, such as mediumship, séances, and channeling. These practices raise intriguing questions about the possibility of communicating with the deceased and provide insights into the nature of consciousness and energy.

A more empirical standpoint is essential when analyzing scientific research and theories on consciousness and the possibility of an afterlife. We will explore the scientific perspectives on these topics, considering the evidence and arguments put forth by researchers and scholars.

Cultural and historical views on the afterlife reveal fascinating variations in beliefs and interpretations. Different cultures and historical periods have provided diverse answers to the question of what happens when we die. We will uncover these interpretations and analyze their cultural and historical significance.

Furthermore, we will investigate the impact of our fear of death on our beliefs and behaviors. The fear of death can shape our decisions, influence our religious or spiritual beliefs, and even drive our pursuit of immortality through legacies and accomplishments.

Lastly, we will consider the viewpoints of skeptics and atheists who reject the idea of an afterlife. Their arguments against life after death force us to critically evaluate our own beliefs and challenge our assumptions.

In conclusion, the fear of death and existential anxiety are fundamental aspects of the human experience. By exploring various perspectives on what happens when we die, we gain a deeper understanding of our own beliefs, values, and the meaning we assign to life.

Death Denial and Coping Mechanisms

Death is a topic that has fascinated and perplexed humanity for centuries. It is a concept that is both universal and deeply personal, as each individual grapples with the inevitability of their own mortality. In the face of this existential dilemma, humans have developed a myriad of coping mechanisms and strategies to deal with the reality of death.

One common coping mechanism is death denial, the psychological defense mechanism that allows individuals to avoid the anxiety and fear associated with their own mortality. Death denial can take many forms, from avoiding conversations about death to engaging in activities that distract from thoughts of mortality. Some individuals may even go to great lengths to preserve their youth and physical appearance, as if to deny the reality of aging and death.

Another coping mechanism that has gained significant attention is the phenomenon of near-death experiences (NDEs). NDEs are often described as profound, transformative experiences that occur when an individual is on the brink of death or has been revived after being clinically dead. These experiences often involve a sense of floating outside one's body, moving through a tunnel, encountering deceased loved ones, and experiencing a profound sense of peace and love. While skeptics argue that NDEs can be explained by physiological and psychological factors, others believe that they provide compelling evidence for the existence of an afterlife.

Reincarnation is yet another coping mechanism that has captivated the human imagination for centuries. The belief in the cycle of birth, death, and rebirth is found in various religious and spiritual traditions around the world. Case studies and research on past-life memories have been conducted, aiming to provide scientific evidence for the existence of reincarnation. These studies often involve children who claim to remember details of past lives that they could not have known otherwise.

Spiritual beliefs also play a significant role in how individuals cope with death. Different religious and spiritual perspectives offer diverse explanations for what happens after we die. Concepts such as heaven, hell, and purgatory provide comfort and hope for believers, offering the promise of an eternal afterlife or the opportunity for redemption.

For those seeking to connect with the deceased, practices such as mediumship, séances, and channeling offer a way to communicate with the spirit world. While these practices are often met with skepticism and controversy, proponents argue that they provide a means of finding solace and closure after the loss of a loved one.

Scientific perspectives on the afterlife approach the topic from a more empirical standpoint, exploring research and theories on consciousness, energy, and the possibility of an afterlife. From the study of near-death experiences to investigations into the nature of consciousness, scientists continue to explore the boundaries of what we know about life after death.

Cultural and historical views on death also shed light on how different societies have interpreted and answered the question of what happens when we die. From ancient Egyptian beliefs in the afterlife to the Tibetan Buddhist concept of the Bardo, cultural perspectives offer a rich tapestry of interpretations and rituals surrounding death.

Death rituals and customs vary greatly across cultures and can provide insights into different cultural perspectives on the afterlife. From elaborate funeral ceremonies to unique burial practices, these rituals serve to honor the deceased and provide comfort to the living.

The fear of death is a powerful force that influences our beliefs and behaviors. Examining the psychological and philosophical implications of this fear can provide valuable insights into the human experience and our understanding of mortality.

Skeptics and atheists, on the other hand, reject the idea of an afterlife altogether. Their arguments against the existence of life after death often center around the lack of empirical evidence and the belief that consciousness is a product of brain activity.

Finally, beliefs about the afterlife have profound ethical and moral implications. They shape our understanding of ethics, morality, and the meaning of life. For some, the belief in an afterlife provides a framework for moral behavior and the pursuit of a meaningful existence.

In conclusion, death denial and coping mechanisms are intricately woven into our human experience. Whether through denial, near-death experiences, reincarnation, spiritual beliefs, afterlife communication, scientific perspectives, cultural and historical views, death rituals, fear, skepticism, or ethical considerations, humans have developed various ways to confront the mystery of death and find solace in the face of mortality. Exploring these coping mechanisms allows us to gain a deeper understanding of our own beliefs and behaviors, as well as the rich tapestry of human experiences surrounding the question of what happens when we die.

Philosophical Perspectives on Death and Fear

In this subchapter, we will delve into the philosophical perspectives on death and fear, exploring the profound questions surrounding what happens when we die and the impact of our fear of death on our beliefs and behaviors.

Death, the ultimate mystery of life, has fascinated and perplexed humans for centuries. Philosophers from various cultural and historical backgrounds have contemplated the nature of death and its implications for human existence. From the ancient Greeks to modern thinkers, philosophical perspectives have shaped our understanding of mortality, the afterlife, and the human condition.

One of the fundamental questions we will explore is the nature of consciousness and its relation to death. Does consciousness cease to exist after death, or does it continue in some form? We will examine the arguments put forth by philosophers such as Plato, who believed in the immortality of the soul, and Epicurus, who argued that death is the end of consciousness.

Fear of death, a universal human experience, also plays a significant role in shaping our beliefs and behaviors. We will analyze the psychological and philosophical implications of this fear, exploring how it influences our understanding of ethics, morality, and the meaning of life. Existentialist thinkers like Søren Kierkegaard and Albert Camus have grappled with the fear of death and its impact on the individual's search for meaning and authenticity.

Furthermore, we will address the viewpoints of skeptics and atheists who reject the idea of an afterlife. We will examine their arguments against the existence of life after death, exploring the implications of their beliefs for ethics, morality, and the search for meaning.

Ultimately, this subchapter aims to provide a comprehensive overview of the philosophical perspectives on death and fear, offering readers a deeper understanding of the profound questions surrounding mortality and the afterlife. By examining different viewpoints and engaging with philosophical arguments, we hope to foster a thoughtful and open-minded exploration of these complex and deeply human concerns.

Whether you are curious about what might happen when we die, interested in near-death experiences and their implications, or seeking to understand the ethical and moral considerations surrounding the afterlife, this subchapter will provide a rich and thought-provoking exploration of the philosophical perspectives on death and fear.

Chapter 10: Skepticism and Atheism: Considering Alternative Viewpoints

Arguments Against the Afterlife

While many people believe in the existence of an afterlife, there are also those who present compelling arguments against this notion. Skeptics and atheists, in particular, have raised several points that challenge the idea of life after death. In this subchapter, we will examine some of these arguments and explore the reasoning behind them.

One of the main arguments against the afterlife is rooted in scientific perspectives. According to skeptics, there is a lack of empirical evidence to support the existence of an afterlife. They argue that consciousness, which is believed to be the essence of our being, is a product of brain activity. When the brain ceases to function at death, consciousness also ceases to exist. From this standpoint, the concept of an afterlife becomes unfounded and without basis.

Another argument comes from the perspective of atheism. Atheists reject the idea of an afterlife primarily because they do not believe in the existence of any deities or higher powers. They view the concept of an afterlife as a mere construct created by religious beliefs to provide comfort in the face of mortality. Atheists argue that our fear of death and desire for immortality are natural, but that does not imply the existence of an afterlife.

Furthermore, skeptics and atheists raise questions about the credibility of near-death experiences (NDEs) and past-life memories. NDEs are often cited as evidence of an afterlife, as individuals claim to have experienced a realm beyond death before being revived. However, critics argue that these experiences can be explained by physiological and psychological factors, such as lack of oxygen to the brain or the brain's attempt to make sense of a traumatic event. Similarly, the purported memories of past lives can be attributed to false memories or suggestibility.

It is also important to consider the ethical and moral implications of belief in the afterlife. Critics argue that the belief in an afterlife can lead to complacency or indifference towards improving the current world. Instead of focusing on making a positive impact in this life, some may become preoccupied with attaining rewards or avoiding punishment in the afterlife. This argument suggests that our beliefs about the afterlife can influence our understanding of ethics, morality, and the meaning of life.

In conclusion, skeptics and atheists present valid arguments against the existence of an afterlife. Their viewpoints challenge the empirical evidence, question the credibility of near-death experiences and past-life memories, and highlight ethical and moral considerations. While belief in the afterlife remains a personal choice, it is important to critically examine these arguments in order to form a well-rounded perspective on this intriguing topic.

Skepticism Towards Near-Death Experiences

Near-death experiences (NDEs) have long fascinated and perplexed both scientists and the general public. These extraordinary accounts of individuals who claim to have glimpsed the afterlife during a brush with death raise profound questions about the nature of consciousness and the possibility of an afterlife. However, despite the abundance of personal testimonies, skepticism towards NDEs remains prevalent.

One of the primary reasons for skepticism towards NDEs lies in their subjective nature. Critics argue that NDEs are merely hallucinations or illusions resulting from physiological and psychological factors. They point to the fact that NDEs often occur during moments of extreme stress or trauma, when the brain may be susceptible to altered states of consciousness.

Furthermore, skeptics argue that the similarities between NDE accounts could be attributed to cultural and societal influences. They contend that individuals may be influenced by popular depictions of the afterlife, such as religious teachings or media portrayals, which shape their experiences and perceptions during near-death situations.

Another source of skepticism towards NDEs stems from the lack of empirical evidence. Despite numerous anecdotal accounts, scientific research on NDEs remains inconclusive. Critics argue that without tangible evidence, such as verifiable information obtained during an out-of-body experience, it is difficult to validate the existence of an afterlife.

Additionally, skeptics highlight the potential for cognitive biases and memory distortions in NDE accounts. They argue that individuals may selectively remember and interpret their experiences based on preexisting beliefs or desires, thereby influencing the content and interpretation of their NDEs.

However, it is important to note that skepticism towards NDEs does not discount the significance and impact these experiences have on individuals. Regardless of the veracity of NDEs, they often provide comfort, reassurance, and a renewed appreciation for life to those who undergo them. Understanding and respecting the subjective nature of these experiences is essential in fostering open dialogue and further exploration.

In conclusion, skepticism towards near-death experiences persists due to their subjective nature, lack of empirical evidence, and the potential for cognitive biases. While skeptics raise valid concerns, it is crucial to approach this topic with an open mind and engage in scientific inquiry to better understand the complex phenomena surrounding the possibility of an afterlife.

Atheistic Perspectives on Life After Death

In this subchapter, we will delve into the atheistic perspectives on life after death. Atheism, as a belief system, rejects the existence of any deities or supernatural entities, including the concept of an afterlife. Atheists argue that there is no scientific evidence to support the existence of an afterlife and that such beliefs are rooted in wishful thinking and superstition.

From a scientific standpoint, atheists often look to neuroscience and psychology to explain the phenomenon of near-death experiences (NDEs), which are often cited as evidence for an afterlife. They propose that NDEs can be attributed to physiological and psychological processes occurring in the brain during moments of extreme stress or trauma. According to this perspective, NDEs are subjective experiences rather than glimpses into an actual afterlife.

Similarly, atheists approach the concept of reincarnation with skepticism. They argue that claims of past-life memories can be explained through psychological factors such as false memories or cultural influences. Atheists contend that there is no verifiable evidence to support the idea that consciousness can survive death and be reborn into a new body.

From an ethical and moral standpoint, atheists often emphasize the importance of living a meaningful and fulfilling life in the here and now. They argue that the belief in an afterlife can undermine the value of this life and lead to a devaluation of human existence. Atheists advocate for a focus on creating a just and compassionate society based on reason and evidence rather than relying on the promise of rewards or punishments in an afterlife.

It is important to note that atheism is a diverse belief system, and not all atheists hold the same views on the topic of life after death. Some atheists may be open to the possibility of an afterlife if scientific evidence were to emerge, while others may remain staunchly skeptical.

By examining the atheistic perspectives on life after death, we gain a deeper understanding of the diversity of beliefs and viewpoints surrounding this fascinating and existential question. It is through exploring different perspectives, including atheistic ones, that we can engage in a more nuanced and comprehensive discussion on the nature of life, death, and the possibility of an afterlife.

Chapter 11: Ethical and Moral Considerations: Exploring the Impact of Afterlife Beliefs

Morality and the Afterlife

When contemplating the possibility of an afterlife, one cannot ignore the ethical and moral implications that arise from such beliefs. The concept of an afterlife has long been intertwined with ideas of reward and punishment, and this has had a profound impact on how societies have shaped their moral codes.

Throughout history, various religious and spiritual perspectives have offered interpretations of the afterlife that directly influence notions of right and wrong. Concepts such as heaven, hell, and purgatory have served as motivations for individuals to lead virtuous lives, with the promise of eternal bliss or eternal torment awaiting them in the hereafter. These beliefs have provided a moral compass for countless individuals, shaping their actions and guiding their decision-making processes.

Additionally, the belief in reincarnation, the cycle of birth, death, and rebirth, has fueled moral considerations in different cultures. The belief that one's actions in this life will determine their circumstances in future lives has led to the cultivation of virtues and the avoidance of harmful behaviors. Case studies and research on past-life memories have further contributed to the exploration of ethical implications associated with the belief in reincarnation.

On the other hand, the possibility of afterlife communication has challenged traditional ethical frameworks. Practices such as mediumship, séances, and channeling have sparked debates regarding the authenticity and morality of attempting to communicate with the deceased. Skeptics argue that such practices exploit vulnerable individuals who are grieving, while proponents argue that they provide comfort and closure.

Scientific perspectives on the afterlife have also offered a unique lens through which to examine moral considerations. The analysis of consciousness and energy from an empirical standpoint has raised questions about the interconnectedness of all living beings and the potential implications for ethical behavior.

Furthermore, cultural and historical views shed light on how different societies have understood and answered moral questions surrounding the afterlife. Death rituals and customs vary greatly across cultures, reflecting diverse beliefs about the journey beyond death and how to honor the deceased.

Ultimately, beliefs about the afterlife have a profound impact on our understanding of ethics, morality, and the meaning of life. Whether one believes in an afterlife or not, the consideration of what happens when we die holds significant weight in shaping individual and collective moral frameworks. Exploring these ethical and moral considerations allows us to delve deeper into the complexities of human existence and the profound questions that arise when contemplating the possibility of life beyond death.

Ethical Dilemmas in the Context of an Afterlife

In the exploration of the possibility of an afterlife, one cannot ignore the ethical dilemmas that arise when considering the implications of such a realm. Beliefs about the afterlife have profound effects on our understanding of ethics, morality, and the meaning of life. This subchapter delves into the ethical considerations that emerge when contemplating what happens when we die.

One of the primary ethical dilemmas is the question of how our actions in this life may affect our fate in the afterlife. Various religious and spiritual perspectives posit that our behavior on Earth determines our eternal destiny. This raises profound questions about justice, accountability, and the concept of divine punishment or reward. Are individuals truly responsible for their actions if they are to be judged in the afterlife? What about those who escape earthly justice? These dilemmas challenge our understanding of fairness and the nature of divine justice.

Another ethical consideration is the impact of belief in the afterlife on our behavior in the present life. Some argue that a belief in an afterlife with rewards or punishments can lead to moral behavior, as individuals strive to secure a positive outcome in the afterlife. However, others contend that such beliefs can also lead to complacency or a lack of accountability, as individuals rely solely on the hope of redemption in the afterlife rather than taking responsibility for their actions in the here and now.

Additionally, ethical dilemmas arise in the realm of afterlife communication. Practices such as mediumship, séances, and channeling have long been used to attempt communication with the deceased. However, questions of authenticity and exploitation are inherent in these practices. Are mediums genuinely connecting with the deceased, or are they preying on the grief-stricken and vulnerable? This raises concerns about the ethical implications of exploiting people's beliefs in the afterlife for personal gain.

Furthermore, the fear of death and its influence on our beliefs and behaviors also present ethical dilemmas. If a belief in the afterlife alleviates our fear, is it ethical to promote such beliefs even if they may lack empirical evidence? Conversely, if an individual's fear of death leads to negative consequences, such as avoidance of necessary medical procedures, how do we ethically address and navigate these situations?

Ultimately, the ethical considerations surrounding the afterlife are complex and multifaceted. They challenge our understanding of justice, accountability, morality, and the impact of beliefs on our behavior. Exploring these dilemmas can provide valuable insights into the intricate relationship between the afterlife and our ethical framework, shedding light on the profound implications of our beliefs about what happens when we die.

Finding Meaning in Life Through Afterlife Beliefs

The concept of an afterlife has fascinated humanity for centuries, providing solace, hope, and a sense of purpose in the face of mortality. In this subchapter, we will delve into the various ways in which individuals and cultures seek meaning in life through their beliefs about the afterlife.

Near-death experiences (NDEs) have been reported by individuals who have come close to death and returned. These accounts often include vivid descriptions of a realm beyond our physical existence, offering glimpses into what may await us after death. We will explore the theories surrounding NDEs and their potential implications for understanding the afterlife.

Reincarnation, the belief in the cycle of birth, death, and rebirth, is another perspective on the afterlife that has captivated millions across cultures. Through intriguing case studies and research on past-life memories, we will examine the evidence supporting this belief and its impact on our understanding of the purpose of life.

Religious and spiritual perspectives on the afterlife provide a rich tapestry of beliefs, ranging from concepts of heaven, hell, and purgatory to ideas of spiritual evolution and liberation. By examining these diverse perspectives, we can gain insight into how different cultures and faiths shape their understanding of what lies beyond death.

Afterlife communication, through practices such as mediumship, séances, and channeling, offers the possibility of connecting with the deceased. We will explore the experiences, testimonies, and controversies surrounding these practices to evaluate their potential for providing comfort and closure to those grieving the loss of a loved one.

Taking a more empirical standpoint, we will analyze scientific theories on consciousness, energy, and the possibility of an afterlife. By examining the latest research and groundbreaking studies, we can gain a deeper understanding of the scientific perspectives on this age-old question.

Moreover, we will investigate how different cultures and historical periods have interpreted and answered the question of what happens when we die. By exploring the rituals, burial practices, and funeral traditions from around the world, we can shed light on the cultural perspectives on the afterlife and its significance in various societies.

Additionally, we will examine the psychological and philosophical implications of humanity's fear of death. By understanding how our fear of mortality influences our beliefs and behaviors, we can gain insight into the profound impact that thoughts of an afterlife have on shaping our lives.

Lastly, we will consider the viewpoints of skeptics and atheists who reject the idea of an afterlife. By examining their arguments against the existence of life after death, we can engage in a balanced and critical discussion that challenges our own beliefs and encourages a deeper exploration of the meaning of life.

Ultimately, by exploring the diverse perspectives on the afterlife, we can gain a deeper understanding of how beliefs about the beyond impact our understanding of ethics, morality, and the meaning of life. Whether we find solace in religious teachings, scientific research, or personal experiences, the search for meaning in life through afterlife beliefs is a deeply human endeavor that has shaped cultures and individuals throughout history.

Conclusion: Reflecting on the Investigation of the Possibility of an Afterlife

Throughout this book, "Science and the Beyond: Investigating the Possibility of an Afterlife," we have delved into a wide range of topics surrounding the question of what happens when we die. From near-death experiences to reincarnation, spiritual beliefs to afterlife communication, scientific perspectives to cultural and historical views, death rituals to fear of death, skepticism to ethical considerations - we have explored various angles to shed light on this age-old mystery.

Our journey into the investigation of the possibility of an afterlife has been fascinating and thought-provoking. We have examined scientific research and theories, studied the accounts and theories surrounding near-death experiences, and delved into cultural and historical interpretations. We have explored the beliefs of different religions and spiritual traditions, as well as the practices of mediumship and afterlife communication.

While the question of what happens when we die still remains largely unanswered, our exploration has provided us with valuable insights and perspectives. We have learned that near-death experiences can offer glimpses into a realm beyond our physical existence, and that many individuals claim to have memories of past lives. We have also discovered that various cultures and historical periods have their own unique interpretations of the afterlife, each shedding light on the complexities of this topic.

Furthermore, we have considered the viewpoints of skeptics and atheists who reject the idea of an afterlife, recognizing the importance of diverse perspectives in any investigation. We have also explored how beliefs about the afterlife impact our understanding of ethics, morality, and the meaning of life.

Ultimately, this investigation has shown us that the possibility of an afterlife is a deeply personal and subjective matter. While science may provide empirical evidence and cultural beliefs may offer guidance, the truth remains elusive. The question of what happens when we die is one that each individual must grapple with in their own way.

As we conclude this exploration, it is important to remember that the search for answers is a journey, and our understanding of the afterlife is continually evolving. By engaging in open-minded dialogue and continuing to explore this topic, we can expand our knowledge and perhaps come closer to unraveling the mysteries of life and death.

In the end, regardless of our individual beliefs or lack thereof, it is the discussions and investigations surrounding the possibility of an afterlife that enrich our lives and deepen our understanding of what it means to be human.